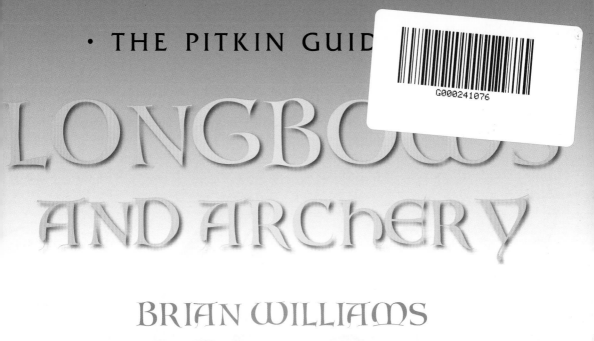

· THE PITKIN GUIDE

LONGBOWS AND ARCHERY

BRIAN WILLIAMS

The longbow was a feared battlefield weapon, especially in western Europe. For 500 years during the Middle Ages, the mighty bow was an icon of military might in England. Skilled archers could bring down knights in armour and their horses. A longbow shot faster than its rival, the crossbow. Archers gave an army battlefield mobility and fearsome firepower. An 'arrowstorm' shot by hundreds of archers could decimate a far greater number of knights, while a lone bowman was a deadly sniper, guerrilla fighter and hunter. Stories spun around bows and bowmen, such as the tales of England's most famous archer, Robin Hood, blend fact and fantasy into the legend of the longbow.

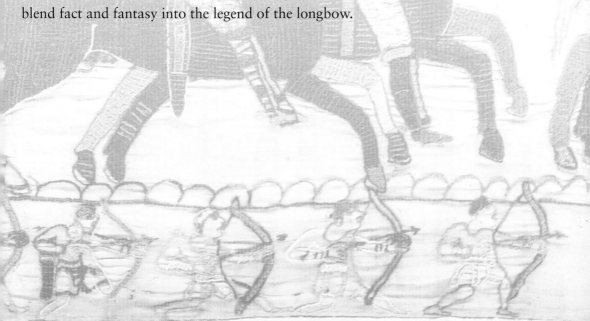

Over 50,000 years ago, a Stone Age hunter made the first bow. He or she bent a length of springy wood strung with a cord of animal sinew or twisted plant fibres, and shot an arrow. The bow was a giant leap in technology. Prehistoric hunters could kill animals as large as mammoths, from a safe distance. Warrior-archers could kill enemies in battle or lie in ambush, killing with silent stealth. The bow remained the supreme missile weapon even after guns were invented more than 500 years ago, and was used all over the world.

BOWS IN THE MIDDLE AGES

In Europe, the longbow was the most feared weapon during the Middle Ages, roughly AD 500 to 1500. Three styles of bow were common: crossbow, short bow and longbow. In Britain, archers preferred the longbow, and with it they won battles right up to the 1500s.

▲ *A medieval hunter with a short bow, from a painted tile in the Chapter House of Westminster Abbey, c.1255.*

▲ *On a 1603 map of South America, a Spanish soldier confronts a giant Patagonian archer.*

The bowman in his cap and leather jacket, with his longbow and sheaf of arrows, won startling English victories in France at the battles of Crécy and Agincourt. Though he seemed no match for knights in armour on charging horses, the archer had an edge. The knight had lance, sword, mace or axe, but the bowman had his longbow, and its iron-tipped arrows.

CROSSBOW

The crossbow was a mini-catapult, slow to load but powerful. The bow was fixed to a wooden stock, with a foot-rest or stirrup for the bowman to brace himself as he drew back the string, which hooked onto a trigger mechanism. Some crossbows had crank handles to wind back the string. Releasing the trigger shot a short arrow called a bolt or 'quarrel'.

Japanese Longbows

The Japanese shot very long bows, some over 8 feet (2.4 metres). These bows had the hand grip two-thirds down, not in the middle. The great bow of Minamoto no Tametomo, a 12th-century Japanese samurai warrior, was said to need three ordinary men to draw it.

> A Japanese archer with longbow; a woodcut by Yanagawa Shigenobu (1787–1832).

How a Bow Works

The bow looks simple: a length or 'stave' of wood, with a string looped about each end. Pulling back the string bends the bow. This 'draw' requires a lot of muscle, not just in the arms, but the whole body. The bow's secret is that it stores muscle energy, and transmits it to the arrow. When the archer lets go or 'looses' the bent bow, the energy is released, sending the arrow whizzing away with great force.

A longbow can be made from a single strip of wood. Composite bows are made from different materials such as strips of wood, horn and sinew glued together. The modern compound bow is even more complex, its sections linked by levers and pulleys. The longbow is straight. Many bows were 'recurved' (like Mongol bows in Asia), with tips that curved away from the bowman. At 6–7 feet (1.8–2.1 metres) the longbow is long, taller than most archers, and so best shot from a standing position. Short bows (typically 4 feet/ 1.2 metres) were useful for hunting and for shooting from horseback.

The crossbow, a very different weapon, was widely used as battlefield light artillery. But it was slow. In Britain, the longbow was first choice – simple, reliable, quick-shooting and deadly.

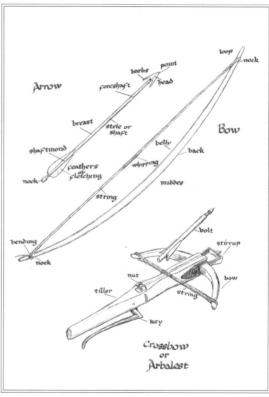

Arrow — barbs, point, foreshaft, head
breast, stele or shaft
shaftmond
feathers or fletching
nock

loop, nock
Bow
belly, back
whipping
middes
string

bending
nock

Crossbow or Arbalest — bolt, stirrup, nut, bow, tiller, string, key

⌃ The key elements of the longbow and crossbow. Crossbows were made by 'arbalastiers' (from the French arbalest), whose craft was distinct from that of the longbow-maker or 'bowyer'.

The bow was a favourite weapon in Ancient Egypt over 4,000 years ago, as we can see from pictures, carvings and tomb-models of archers. Archer-kings rode to battle in chariots. The Hittites, Assyrians, Persians and Chinese also had chariot archers, to dash around the battlefield.

Hunters all over the world used the bow, often tipping the arrows with poison. No animal was safe. Liangulu hunters in Kenya shot elephants with bows and arrows, and Native Americans riding across the Great Plains hunted buffalo.

◀ *A 7th-century carving from a Hindu temple in India shows the god Siva drawing a longbow.*

HORSE-ARCHERS

The Ancient Greeks, who admired athletic grace and warrior-heroes, shot bows and pictured the gods Apollo and Artemis as archers. The Roman legions preferred the javelin and sword, and the Roman army often turned to foreign soldiers for its archers. Roman soldiers were superbly trained, but archers could be their undoing. Bowmen dealt the Roman army one of its worst defeats in 53 BC, when Parthian horse-archers destroyed Roman legions at the Battle of Carrhae, in what is now Turkey. The Parthians used composite bows that shot arrows over 300 metres, and from horseback shot even while galloping away, a trick that passed into history as the 'Parthian [parting] shot'. Hundreds of camels supplied the archers with fresh arrows

The 'parting shot', from an Ottoman Turk horse-archer. From a 15th-century miniature, in the Topkapi Palace Museum, Istanbul.

The archer kept his arrows (usually 12–24) in a pouch or quiver. In battle, he stuck arrows through his belt or in the ground. To shoot, he first 'nocked' the arrow onto the string. Then he 'drew', bending and raising the bow, using his whole body strength. After sighting (taking aim) he swiftly 'loosed' to release the arrow, then reloaded. A good archer could shoot 10 arrows a minute, or faster.

◄ *An archer in armour, illustrated in the Stuttgart Psalter, a book of Psalms dating from about 830.*

to maintain relentless attacks, which, with armoured cavalry charges, left thousands of Roman soldiers dead.

Longbows in Europe

In Europe, the bow was used long before Roman times. In Scandinavia, archaeologists have found bows in ship burials, showing that a good bow was a valued item. Britain's oldest longbow is a yew bow from Meare Heath in Somerset, made in around 2690 BC. About 6 feet (1.8 metres) long, it would have looked familiar to a medieval archer.

In the Middle Ages, many thousands of bows must have been made: short bows (probably more common until the AD 1300s) and longbows. Yet no complete English medieval longbows survive. Old bows were burned as firewood or rotted away on rubbish tips. Only fragments turn up, like the bow found in the moat at Berkhamsted Castle in 1931 (a crossbow?) and an ash stave found at Southampton in 1966 (part of a bow?). A bow found at Waterford in Ireland, dating

between 1150 and 1250, looks like a Norman-style short bow. Arrows and rusty arrowheads are more plentiful. So all the more remarkable is the collection of 137 Tudor longbows, recovered from Henry VIII's warship *Mary Rose* (see page 21).

➤ *Longbow shooting technique, showing feet and finger positions. The archer 'holds', or pauses, briefly before 'loosing' the arrow.*

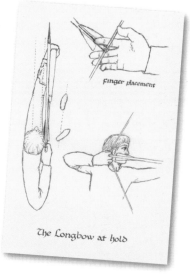

finger placement

The Longbow at hold

In medieval battles, such as Crécy and Agincourt (see pages 16–19), hundreds of archers loosed arrows in volleys. This 'arrowstorm' rained down on men and horses. Archers could keep shooting in this way for only a few minutes before they used up their arrows, and their strength. It was usually enough. If they ran out of ammunition, and the enemy was still advancing, they took to their heels.

THE BOW IN BRITAIN

From the AD 700s, Viking sea-raiders ravaged, and settled, the British Isles.
Like the Anglo-Saxons, the Vikings (from Scandinavia) liked to fight with spear, sword and axe. They used bows too, in ship fights and when hunting. When Danish Vikings slew King Edmund of East Anglia, they are said to have peppered the unfortunate king with arrows.

ANGLO-SAXON ARCHERS

Alfred the Great led the English fight against the Vikings. Most of Alfred's army were part-time soldiers, from the fyrd or militia. The king's bodyguard had armour, shields, swords and great axes, but fyrd soldiers brought whatever weapons they could find, including bows. English and Vikings usually fought on foot, hand-to-hand, with archers firing from a distance, trying to keep out of the way of the flailing axes and swords.

In 1066, England's last Anglo-Saxon king Harold II faced two invading armies in three weeks. On 25 September 1066 the English took on Norwegian Vikings. At the Battle of Stamford Bridge in Yorkshire, bowmen were in the thick of the bloodshed. The Viking king Harold Hardrada was killed by an arrow to the throat. The death of a leader could turn a battle, and so it was to prove at the Battle of Hastings weeks later.

▲ *An English warrior hit by an arrow at Hastings in 1066. Tradition says Harold was struck in the eye, though it is as likely he was hacked down by swords and axes. The Latin text simply says, 'King Harold was slain'.*

ARROW WOUNDS

Many soldiers at Hastings must have died from arrow wounds, like the warrior shown in the Bayeux Tapestry clutching his face. A war arrow was not easy to pull out. If the arrowhead stayed in the flesh, the wound might turn septic. Medieval surgeons often chose to push the arrow right through, and out the other side.

▼ *The Battle of Hastings. In this scene from the Bayeux Tapestry, Norman knights are supported by archers shooting high so arrows fall on the English battle line.*

Archers at Hastings

As he celebrated victory, King Harold received news that Duke William and his Norman army had crossed the Channel and landed in Sussex. To deal with this challenge, the English king raced south, gathering men on the way. The English and Normans clashed on 14 October 1066, near Hastings. The English fought in their old way, standing their ground, soaking up attacks all day long. The Normans used more varied tactics: their knights charged on horses, fake retreats lured the English into breaking ranks, and all the time archers shot at the English battle line. The Bayeux Tapestry, a picture-strip account of the Norman Conquest, shows archers, almost all looking like Normans. Many English bowmen had perhaps gone home to their farms after Stamford Bridge, or been left behind in Harold's hectic dash to meet his fate.

◀ This 1975 embroidery, by Sybil Andrews, shows Edmund of East Anglia, English king and martyr, killed by Viking arrows in 869/870.

Arrows hit the English from the front and from above, and battered by Norman horse charges, their valiant 'shield-wall' finally broke. Harold was killed – some historians suggest not shot in the eye by an arrow, but hacked to death by Norman knights.

Bristling with Arrows

In 1138 King David I led a Scottish army into England. Norman-English and Scots fought in Yorkshire, around a cart bearing the English standard or flag. In this 'Battle of the Standard' Scots spearmen fell to English arrows. The chronicler (historian) Ailred of Rievaulx described the archers' victims as looking 'like a hedgehog, bristling all round with arrows'.

◀ Traditional illustrations of Harold's death at Hastings show the king hit in the eye, as Norman arrows showered down on the English.

▲ *In this German illustration of the hunting year, about 1480, a poor man with a crossbow helps himself to a rabbit, unseen by the nobleman on horseback.*

Bowmaking is a skilled craft. Welsh and English bowmakers (bowyers) agreed that the best longbows were made from the sound trunk of a yew tree, though other woods could be used, such as elm, ash, birch, oak, blackthorn, beech, elder and hornbeam. Mature trunk wood was best. Sapling and branches were fit only for inferior bows and children's toys. The tree was cut down in winter, and the trunk split with wedges into staves (bow-lengths), which were then left to season for up to five years.

➤ *A modern Tudor-style yew longbow, 82 inches (208 centimetres) long with a draw weight of 115 pounds (52 kilograms).*

SHAPING THE BOW

To make the bow, the bowyer first shaped the back (the side facing the target), then the sides and lastly the belly (facing the archer). He usually 'stacked' the belly, giving it a round or D-shape cross-section. At the end of each limb, he cut a notch for the bowstring. Many bows had a horn nock, glued to the limb-tip. To fit the string, he braced the bow

ARROWHEADS

Medieval arrowheads were usually made of iron. Hunting arrows had broad heads, some curved to slice off the heads or wings of birds. War arrows had forked, barbed or broad heads, to inflict nasty wounds. The long pointed bodkin could pierce armour.

Bow weights

The force needed to draw a bow is known as its weight. Modern tests suggest 80–150 pounds (36–68 kilograms) for longbows. Few modern bows exceed 60 pounds (27 kilograms). The medieval archer was very strong.

◀ *Victims of medieval massacre: Viking raids, border wars, religious conflicts and blood-feuds ravaged early medieval Europe. The bow was a deadly 'hit and run' weapon.*

against his leg and looped the bowstring into the nocks using a 'bowyer's knot' (a timber hitch) for the lower loop. The bow was then 'armed' or 'bent up'. Bows could snap, so the bowyer 'taught the bow to bend', gradually flexing it by hand or holding the bow in a wooden rack and hanging weights from the string. The ideal shape was a perfect arc.

To ready a bow for a long life, it could be treated with linseed oil and beeswax. The handle (centre grip) was bound with cord or leather.

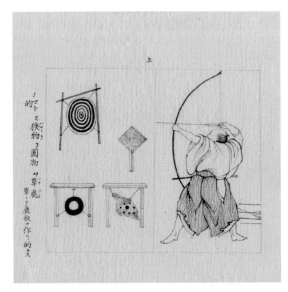

▲ *A Japanese archer draws a longbow and is about to shoot, with several types of target, 1878.*

Cord might also be 'whipped' or wound around the mid-point of the string, where the arrow was 'nocked'.

Shooting

Many archers wore a protective wrist strap or brace on their bow hand. To shoot, the archer stood sideways on. He 'nocked' the arrow into the string, and raised the bow as he bent it, using all his body strength (not just his arms). Children often hold arrow and string between finger and thumb, but the English longbow style was three fingers. At full draw, the archer's bow arm was stretched, his other arm bent to hold the string close to his cheek. Then he 'loosed' the arrow.

Strings and Arrows

Bowstrings were made of flax or hemp, though animal sinew and silk were also used. Arrows were made of wood, usually trimmed to a slim rounded shaft. The longbow arrow was 28–30 inches (71–76 centimetres) long, though some arrows were longer. At the bowstring end or 'shaftment' was a V-cut or 'nock' for the string. Trimmed feathers, or fletching, were glued to the shaftment. These 'vanes' keep an arrow steady in flight. Grey or white goose feathers were an English fletcher's first choice, sometimes dyed red, blue or black.

BOWMEN IN BATTLE

In medieval warfare, archers often took part in a battle for a castle. This usually meant a siege. The attackers surrounded the castle, tried to bash down or climb over the walls, and attempted to starve the defenders out. Archers inside the castle shot from its battlements, taking cover behind merlons (the raised bits). From towers they shot through slit-windows or arrow-loops, made wider on the inside so the archer had a good field of fire. He could shoot down through a fishtail opening at the bottom of the loop, and share a large cross-loop with another archer.

Siege Attack

The ground around a castle was kept clear of trees, so enemy bowmen could not creep up and let loose unseen. To attack across open ground, archers sheltered behind mantlets (wooden shields on wheels) or they carried pavises (large shields with spikes to push into the ground), from behind which an archer could shoot. Siege towers or belfreys were pushed up against the castle walls. The towers' wooden sides were hung with water-soaked cow-hides for protection from fire-arrows and blazing oil from the castle. Inside were assault troops and archers. The archers' arrows kept wall-defenders pinned down, as the attackers stormed the battlements. Once inside the castle, archers joined hand-to-hand battles using swords and knives, or even their bows as clubs.

▲ *A sea battle, from the 1400s. Ships grappled close together as archers joined in the hand-to-hand fighting.*

Logistics

Sieges called for huge effort, and a lot of arrows. To attack Kenilworth Castle in 1266, King Henry III ordered 300 sheaves of arrows from Surrey and Sussex (a sheaf usually comprised 24 arrows). In 1304 his son Edward I, laying siege to Stirling Castle in Scotland, ordered 130 bows all the way from London, and more from Yorkshire and Newcastle. Four bowmakers were with Edward's army, to keep up the archers' firepower.

Archers cost less than knights. Records show that in 1269 archers at Carlisle Castle were paid 2*d* (2 old pence) a day. A knight got 24*d* (twelve times as much). Longbowmen ranked with road-menders and tree-choppers. Crossbowmen were paid more: the same as skilled workers.

Power Shooting

At the siege of Abergavenny Castle in 1182, Welsh archers shot arrows through an oak door 'a hand thick' (4 inches/10 centimetres?), according to the chronicler Gerald the Welshman. He also tells how a Welsh bowman shot an English rider through the thigh. The arrow pierced mail armour, leather saddle, then went into the horse!

THE CLOTHYARD SHAFT

The longbow arrow was called the 'clothyard shaft'. The clothyard was a measure, for cloth, and, at 37 inches (94 centimetres), slightly longer than the common 36-inch yard.

▲ *In a siege, longbows and crossbows were used as mobile artillery for both attack and defence.*

BORDER WARS

In his conquest of Wales in the late 1200s, Edward I's English army fought against Welsh archers, whose skill was much feared. Welsh bowmen later joined English armies to fight in Scotland and France, and archers formed the main part of castle garrisons. Edward built new castles in Wales, and took over Welsh castles, like Dryslwyn, in Carmarthenshire. Its garrison records show it was defended by 2 knights, 22 men at arms, 20 crossbowmen and 80 archers – half of them Welsh. Bowmen were issued with a bow and 12 arrows. A longbow cost only 2s (24 old pence), compared with 5s (60 old pence) for a crossbow.

⌃ Robert the Bruce statue at Stirling Castle. At Bannockburn, the outnumbered Scots had few archers (from the Ettrick Forest), but Bruce's cavalry drove the English bowmen from the field.

KILLING THE LEADER

A king or a noble knight in splendid armour on a magnificent horse stood out on the battlefield. He made a tempting target for a bowman. At the Battle of Shrewsbury in 1403, the young Prince Hal (later Henry V) fought Henry Percy (known as 'Hotspur'). Son of the rebel Duke of Northumberland, Hotspur was England's most famous warrior-knight. He was killed by an arrow, and his family's rebellion was ended.

At the Battle of Falkirk, volleys of arrows from English archers devastated close-ranked Scottish spearmen.

king Robert the Bruce splitting the skull of an English knight in single combat. The English knights charged uphill, but stumbled into pits dug by the Scots, and got hopelessly squashed together and bogged down. Both armies had archers, but for once the longbow was on the losing side. English archers did not shoot much, for fear of hitting their own side, and were driven off by Scots cavalry. Robert the Bruce won a famous victory.

Archer Tactics

In most battles, spearmen on foot and knights on horses were very vulnerable to archers. In 1333, at Halidon Hill near Berwick, King Edward III defeated a Scots army led by Sir Archibald Douglas, and the English tactics served them well for the next 100 years. Edward's men at arms and knights fought on foot, in groups called 'battles'. His archers were placed between the battles, and to either side, in wedge formations. Volleys of longbow arrows hit the Scots, from the front and sides, and as bravely as they attacked they could not withstand the barrage. Many Scots soldiers were felled by swords, spears and axes, while archers picked off those who tried to escape. Thousands of Scots were killed.

Bows in Scotland

In the English-Scottish wars of the Middle Ages, both sides used archers. At the Battle of Falkirk in 1298, English longbowmen and horsemen proved too much for the Scottish 'schiltrons'. The schiltrons were blocks of foot-soldiers with pikes (very long spears). The Scots marched forward, but were assailed by arrows. Abandoned by their own knights, they were then killed or scattered by English cavalry.

It was a different story at Bannockburn in 1314. This famous battle began with the Scots

Chaucer's Bowman

Medieval poet Geoffrey Chaucer has a yeoman-archer among the pilgrims in his *Canterbury Tales*:

Clad in coat and hood of green, a sheaf of peacock arrows bright and keen under his belt.
His arrows had no draggled [bedraggled] feathers, And in his hand he bore a mighty bow.

The bow was the outlaw's ideal weapon: powerful, silent, deadly at long range. No wonder England's most famous legendary outlaw, Robin Hood, was a master bowman. Robin Hood stories and films feature all the tricks of the archer's trade: fire arrows, whistling arrows, arrows with messages, arrows trailing cords for daring escapes from castles. Outlaws dressed in 'Lincoln green' (a woollen cloth) roam Sherwood Forest, outwit wicked foes, fight bravely, shoot brilliantly and rob the rich only to give to the poor. The dying Robin shoots his last arrow to mark his grave.

The Robin Hood legends made the longbow a hero's weapon. In real life, bows and arrows were used in robbery and murder. People were shot in tavern brawls or ambushed in lonely lanes.

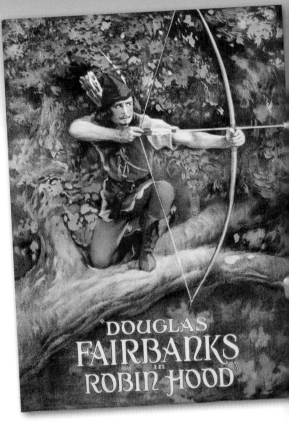

"DOUGLAS FAIRBANKS in ROBIN HOOD"

∧ *Hollywood was quick to realise the appeal of Robin Hood, played by action star Douglas Fairbanks in this 1922 silent film.*

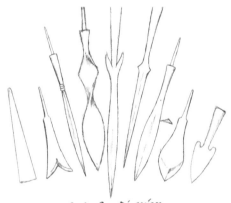

Early Scandinavian and Viking arrowheads

< *Arrowheads used by Viking and early medieval archers. Broad blades caused severe wounds; slender points could pierce armour.*

Early mediaeval arrowheads

STUNT SHOOTING

In Hollywood films, archers perform amazing feats. They shoot through a hangman's rope on a gallows. They shoot one arrow to split another. Some of these stunts are faked, but others are genuine trick shots by experts with the bow.

Poachers in deer forests fought battles with sheriffs and gamekeepers. Villagers and townsfolk kept bows at home, and were expected to join the 'hue and cry' (a posse) to catch a lawbreaker.

Accident or Murder?

Even archery practice could be dangerous, although coroners' records of accidental deaths suggest that in the 1300s football was more dangerous than archery, and staggering home drunk at night considerably more risky! Yet with bows twanging in dense forests, hunting accidents were not uncommon. Some look suspicious. In 1100

Tall Tales

Epic tales from India describe archers shooting four arrows from one bow – and hitting four targets! Archers bring down war elephants in battle – like Legolas, the elf-archer in the *Lord of the Rings* films – and in one *Rambo* movie, the hero downs a helicopter with an arrow.

King William II (William Rufus) was shot dead while hunting deer in the New Forest. The archer Walter Tirel fled. William's brother promptly seized his brother's gold and crown, to become Henry I.

Kings needed bowmen for their army, and in 1363 Edward III decided too many men were neglecting their bows. He ordered all fit men to 'learn and exercise the art of shooting', to practise archery on Sundays and not to waste energy on 'throwing stones, hand-ball, football or cock-fighting'.

Murder by Longbow

Medieval law clerks carefully recorded details of arrow crimes. In 1274, Wymund le Chanu from Colmworth in Bedfordshire shot Hugh Bel in the thigh. The barbed arrow made a severe wound, and the victim died, probably from loss of blood. In 1298, in Leicestershire, John of Tylton shot Simon of Skeffington. The arrow tore a hole in Simon's chest 6 inches (15 centimetres) deep.

◁ *This statue of Robin Hood stands near Nottingham Castle.*

15

Wars in France drew archers from all over Britain. In 1345, 125 bowmen left Staffordshire, 100 from Shropshire, others from Wales. With them went shiploads of stores: salted meat, bacon, cheese, bread, oatcakes, peas, beans and dried herrings, thousands of arrows bound in sheaves and bows wrapped in canvas.

England's wars with France, known as the Hundred Years War, produced some of the most famous bowmen battles – Crécy (1346), Poitiers (1356) and Agincourt (1415). Archers shooting from ships also helped the English win the sea Battle of Sluys (1340).

THE BATTLE OF CRÉCY

At Crécy, the French outnumbered the English army of Edward III. Fighting began in the late afternoon, in a thunderstorm. Edward's archers had snatched a meal, keeping their bowstrings dry under their hats. When the rain eased, the sun shone, straight into the eyes of the French. The English were pleased: this was what they had hoped for. They could see plenty of targets. Genoese crossbowmen on the French side moved forward to shoot, but were struck down by volleys of longbow arrows. French knights charged through, at least 15 times, but were hit by

Crécy and Poitiers boosted the longbow's formidable reputation. In 1385, at the Battle of Aljubarrota, 100 battle-hardened English archers helped the King of Portugal defeat the armies of Castile and France.

Wagons and pack horses brought up spare arrows. In battle, 100 archers could shoot 1,000 arrows a minute, so an efficient supply train was vital. Archers who ran out of shafts shot back enemy arrows picked up from the ground and pulled from dead bodies.

an English arrowstorm on each occasion. Archers fired into the faces of the enemy, or shot high, so their arrows fell like rain. The French dared not look up, for fear of being hit in the face; if they raised their shields, their bodies would be unprotected. Not even armour saved them. Their only hope was to charge, to get at the archers.

Boys ran to and from English supply wagons, bringing more arrows. In brief lulls between charges, the archers picked up spent shafts from the battlefield. King Edward's son, the teenage Black Prince, was in the thick of the fighting and 'won his spurs'. When the battle was won, exhausted archers slept by fires on the battlefield, amid thousands of dead and dying.

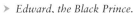

In this re-enactment, the near archer is about to shoot, while the other begins to draw his bow.

Havoc at Poitiers

Poitiers, in 1356, was another archer victory. This time the Black Prince was leading the English, again marauding in France. Needing reinforcements, he ordered more archers, including 500 bowmen from Cheshire, who set sail from Plymouth. The French were thirsting for revenge, and caught up with the English in September. The English deployed on high ground, the archers taking cover behind a hedge and a barricade of trenches and wagons. Hundreds of French horses fell, the bowmen shooting at the animals' sides, which had less armour. Many French knights fought on foot, only to be killed or captured by bowmen, English knights and mounted archers. At Poitiers that day English archers caused 'great havoc', according to the French chronicler Froissart.

Edward, the Black Prince.

Scots Archers

Scottish archers were less famed than Welsh or English bowmen. But King James I of Scotland (murdered in 1437) was an archer and ordered all 'yeomen of the realm between 60 and 16' to have bows and arrows. A proud tradition of Scottish bowmanship lives on in the Royal Company of Archers, the queen's bodyguard in Scotland, founded in 1676.

▲ *The Battle of Agincourt, as shown in the 15th-century St Alban's Chronicle, with the English on the right of the picture.*

The Battle of Agincourt on 25 October 1415 did much to create the myth of the longbow. England's King Henry V, who claimed the French crown, landed in France in August 1415, with an army of about 10,000 men. Of these soldiers, 8,000 were longbow archers, who doubled as all-purpose infantry.

Before the Battle

After a month-long siege, Henry captured the northern French port of Harfleur. Should he march inland to attack Paris and unseat the French monarch? Or head back home, before winter? (Medieval armies usually campaigned only in summer.) Henry challenged the French king's son to single combat, to settle the matter. There was no response. Common sense and older advisers argued for a swift withdrawal: the English army was much reduced by combat, desertion and disease. Henry had only 6,000 soldiers left, but still insisted on a show of strength. He would march through northern France, before taking ship for England.

The French, by now thoroughly enraged, caught up with Henry in October, as the weary

Archers at Agincourt shoot from behind their defensive stakes, as illustrated in Kipling's *Pocket History of England* by C.R.L. Fletcher and Rudyard Kipling.

they could barely swing their swords. Many were crushed underfoot, and they were easy prey for archers who ran to finish them off.

When the battle ended, between 7,000 and 10,000 French lay dead. The English had lost a few hundred at most, including the king's brother, the Duke of York. Wearily, Henry asked the name of a castle nearby. 'Azincourt', he was told, and so the Battle of Agincourt got its name.

English trudged for the coast. The French army was huge: perhaps 40,000 knights and men at arms; 4,000 archers and crossbowmen. Unable to avoid battle, Henry drew up his armoured soldiers in three blocks, flanked by archers. They waited and prayed. Then, knowing if they waited longer they would be too exhausted to fight at all, Henry ordered his archers to attack.

The Archers' Battle

Each archer carried two sharp-pointed wooden stakes. They advanced into bowshot range (about 650–980 feet or 200–300 metres), then hammered in their pointed stakes, which gave some protection from charging knights. They bent their bows, and the sky suddenly darkened with arrows. The first volleys of light arrows with barbed heads hit the French horses. Wounded animals reared, threw their riders and careered through the massed ranks. Now, the English shot heavy arrows, piercing armour of knights so closely packed that

Medieval-style arrows like those used at Agincourt.

TUDOR BOWMEN

By the mid-1400s, the bow was sharing the battles with primitive handguns – noisy, smoky and not very accurate. Still, bowmen played their part in the vicious battles of the civil war in England, known as the Wars of the Roses. At the Battle of Towton in 1461, perhaps the biggest battle ever fought in England, high winds and driving snow helped the Yorkist archers. With the wind at their backs, they outshot the Lancastrian bowmen whose arrows fell short.

By the reign of Henry VIII (1509–47), the longbow was regarded as old-fashioned, but Henry loved the bow. He could outshoot his own bodyguard, and in France at the extravagant royal pageant of 1520, known as the Field of Cloth of Gold, the king showed off by hitting the target at '12 score yards' (240 yards/219 metres). Henry

sent archers to Spain, to fight for his father-in-law, Ferdinand of Aragon. The English disgraced themselves by getting drunk, but impressed the Spanish by their skill with yew bows.

FLODDEN FIELD

The Battle of Flodden, fought in Northumberland in 1513, was the biggest between English and Scots armies. James IV of Scotland invaded England, while Henry VIII (his brother-in-law) was away at war in France. The English army was led by the Earl of Surrey, and the battle began in heavy rain. The Scots charged, but English archers chased them back up a soggy hillside. King James was killed, hit by an arrow to the head, along with perhaps 10,000 Scots in this, the last great longbow battle.

▽ *The Mary Rose at sea. A modern painting by Geoff Hunt reflecting current ideas about how the ship looked. (Geoff Hunt, PPRSMA)*

MARY ROSE BOWS

The *Mary Rose* finds included 137 yew warbows and more than 3,500 arrows. Most of the arrows were made from poplar wood. More than 90 skeletons were found in the wreck. Studies of these remains, and evidence from the ship such as food and clothing remains, tell us much about Tudor seamen and archers.

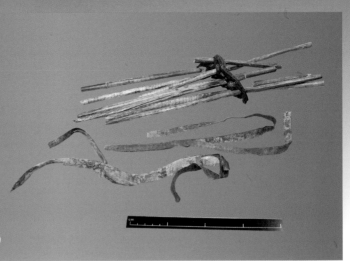

> *Mary Rose arrows, with the spacer in which they were kept. (© Mary Rose Trust)*

GUNS OR ARROWS?

Ships now went to sea with cannon and gunpowder smoke shrouded battlefields. Improved metal armour could stop even longbow arrows. It took less time to train a gunner than a longbow archer. So who wanted bowmen?

Henry VIII did. In 1528 he condemned the 'newfangleness and wanton pleasure that men now have in using crossbows and handguns'. In the 1530s longbows were still being stored in the Tower of London and fletchers paid to repair 'fectyff' (worn-out) arrows. We know from the wreck of the royal ship *Mary Rose*, which sank in 1545 in the Solent and was raised by archaeologists in 1982, that Henry VIII's warships went to sea with elite archers. Archers could sharp-shoot, killing enemy sailors and commanders, as well as quick-fire at close range. Bows were still in the royal armouries at the time of the Spanish Armada in 1588, but their days were numbered.

ARCHERS' STRAINS

Tudor archers were tall, probably around 6 feet (1.8 metres). They had to be strong, with evidence of extra-muscular bow arms. Their skeletons show evidence of repetitive shoulder and back injuries, caused by years of shooting heavy longbows.

< *Archers shoot fire-arrows from Southsea Castle to celebrate the opening of the new Mary Rose Museum in 2013. (Helen Yates Photography)*

By 1627 only four bowyers were at work in the city of London. Fletchers were similarly scarce. Though an archer could shoot further and faster than a musketeer, the tide of war was turning against the bow. In the 1640s, Charles I still had archers in the royal army for the Civil War, but though records show thousands of arrows stockpiled in weapons stores, there were very few bows.

FINAL SHOTS

Charles II enjoyed archery, and it became a genteel pastime for gentlemen and ladies. Bows are also said to have been used in 1688 in a Scottish clan battle between Mackintoshes and MacDonalds. And in 1791 two gentlemen fought a bow duel, each shooting three arrows, 'without damaging each other'.

By the 1800s British soldiers were firing muskets and rifles, but faced arrow-fire from

▲ By the early 1900s archery was gaining popularity as a sport, as this 1920s photograph suggests.

▼ The archery event at the 2012 London Olympics. Archery flourishes still, on the strength of its ancient tradition

Most Deadly ...

Longbow expert and author E.G. Heath suggested that 'the bow and arrow is probably responsible for more casualties in war than any other weapon ...' Even the Second World War commandos occasionally used bows, to kill fast and silently: it's said a longbow victim seldom utters a cry.

archers. During the battle for Lucknow in 1857, an Indian archer's arrow passed right through a British soldier and landed behind him. The soldier 'fell stone dead', according to his sergeant major. As guns improved in the 1800s, the bow was phased out as a war weapon, but was still being used by hunters in the 1900s. Some modern game-hunters still use bows.

Archery For All

In 1829, *The Young Lady's Book* ('a manual of elegant recreation, exercises and pastimes') called archery a healthy and agreeable open-air pastime for ladies. It suggested two sets of targets, so archers could shoot at one, walk up and gather their arrows, and shoot back at the second target. Archery first appeared as an event at the Olympic Games in 1900. Today, top-level archery is an intense, high-tech sport, but many people enjoy archery for fun.

➤ *Today's archers use sophisticated bows for precision shooting. This is England's Nicky Hunt at the 2010 Commonwealth Games in India.*

Last Bow?

Even in the Second World War (1939–45) one or two soldiers refused to give up the bow. Captain Jack Churchill took his bow on patrol with the British army in France in 1939. He let loose at German positions – though at this early 'phoney' stage of the war, British soldiers had been told not to 'provoke' the enemy. By May 1940, German armies were invading Holland, Belgium and France, and this time Churchill used his bow in earnest, shooting a German.

Special Effects Bows

Films such as *Avatar* and *The Hunger Games* have renewed interest in archery, with bow tricks figuring in the special effects action. In *The Hunger Games*, Katniss Everdeen's skill with her hunting bow gives her heroic status, and cinema magic brings the bow icon to a new audience.

Battle Abbey
High St, Battle, Hastings and Battle, East Sussex
TN33 0AD, www.english-heritage.org.uk/1066

Dover Castle, Kent
Castle Hill, Dover, Kent CT16 1HU,
www.english-heritage.org.uk/daysout/properties/
dover-castle/

Imperial War Museum London
IWM London, Lambeth Road, London SE1 6HZ,
www.iwm.org.uk/visits/iwm-london

Imperial War Museum North
The Quays, Trafford Wharf Road, Manchester
M17 1TZ, www.iwm.org.uk/visits/iwm-north

Manchester Museum, University of Manchester
Oxford Road, Manchester M13 9PL,
www.museum.manchester.ac.uk/collection/archery

Mary Rose Museum
No. 3 Dock, Main Road, HM Naval Base,
Portsmouth PO1 3PY, www.maryrose.org

Museum of Archery at Crépy-en-Valois, France
Archery and Valois Museum, Rue Gustave
Chopinet, 60800 Crépy-en-Valois,
www.crepyenvalois.fr/archery_and_valois_
museum.html

The National Army Museum
Royal Hospital Road, Chelsea, London
SW3 4HT, www.nam.ac.uk

National War Museum of Scotland
Edinburgh Castle, Castlehill, Edinburgh,
Midlothian EH1 2NG, www.nms.ac.uk/
our_museums/war_museum.aspx

Nottingham Castle
Castle Place, Nottingham NG1 6EL,
www.nottinghamcity.gov.uk/Castle

Royal Armouries, HM Tower of London
London EC3N 4AB, www.royalarmouries.org/
tower-of-london

Royal Armouries Museum, Leeds
Armouries Dr, Leeds LS10 1LT,
www.royalarmouries.org/leeds

Royal Artillery Museum – Firepower
Royal Arsenal, London SE18 6ST,
www.firepower.org.uk

The Scottish Archery Centre
Unit 15, Fenton Barns Retail Village,
North Berwick, East Lothian EH39 5BW,
www.scottisharcherycentre.co.uk

Wallace Collection
Hertford House, Manchester Square, London
W1U 3BN, www.wallacecollection.org

Warwick Castle
Warwick, Warwickshire CV34 4QU,
www.warwick-castle.co.uk

Windsor Castle
Windsor, Berkshire SL4 1NJ,
www.royalcollection.org.uk/visit/windsorcastle